Wild Weather

HURRICANE

Catherine Chambers

www.heinemann.co.uk/library
Visit our website to find out more information about **Heinemann Library** books.

To order:

 Phone ++44 (0)1865 888066

 Send a fax to ++44 (0)1865 314091

 Visit the Heinemann Bookshop at www.heinemann.co.uk/library to browse our catalogue and order online.

First published in Great Britain by Heinemann Library, Halley Court, Jordan Hill, Oxford OX2 8EJ, a division of Reed Educational and Professional Publishing Ltd. Heinemann is a registered trademark of Reed Educational & Professional Publishing Ltd.

OXFORD MELBOURNE AUCKLAND JOHANNESBURG BLANTYRE
GABORONE IBADAN PORTSMOUTH NH (USA) CHICAGO

Designed by Visual Image
Illustrations by Paul Bale
Originated by Ambassador Litho Ltd.
Printed and bound in South China.

ISBN 0 431 15061 3

06 05 04 03 02
10 9 8 7 6 5 4 3 2 1

British Library Cataloguing in Publication Data

Chambers, Catherine
Hurricane. – (Wild Weather)
1. Hurricanes – Juvenile literature
I. Title
551.5'52
ISBN 0431150613

Acknowledgements

The Publishers would like to thank the following for permission to reproduce photographs: Associated Press pp4, 7, 9, 13, Colorific p17, Corbis pp5, 11, 16, 20, 23, 24, 27, 29, PA Photos p21, Panos p28, Photodisc pp14, 25, Rex Features pp15, 19, 22, 26, Robert Harding Picture Library p12, Science Photo Library p10, Stone p18.

Cover photograph reproduced with permission of Pictor.

The Publishers would like to thank the Met Office for their assistance in the preparation of this book.

Any words appearing in the text in bold, **like this**, are explained in the Glossary.

Contents

What is a hurricane?

A hurricane is a huge storm that builds up over the **oceans**. Strong winds hit the land. Swirling clouds bring heavy rain.

Hurricane winds blow roofs off and smash windows. They snap trees and flatten **crops**. Streets and homes can be flooded by heavy rain or huge waves from the sea.

Where do hurricanes happen?

Hurricanes happen in a region called the **Tropics**. The Tropics are hot because the heat of the Sun is stronger in this region. Hurricanes are called 'typhoons' in some parts of the world.

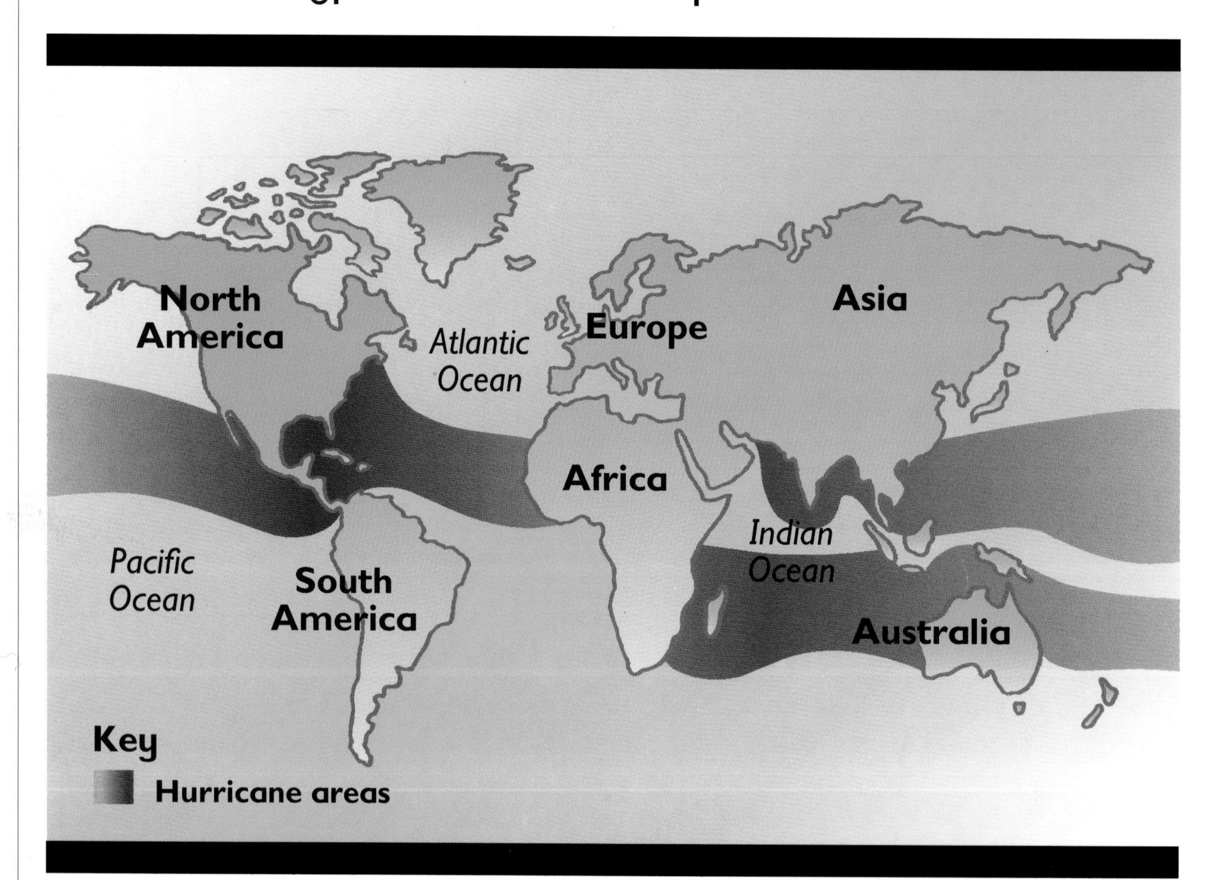

This place lies in the Tropics. Very strong hurricanes happen here because it is close to warm **oceans**. Warm oceans help hurricanes to form.

How do hurricanes form?

Masses of air are always moving. A warm mass of air usually rises. This makes a low **pressure** area. A cold mass sinks. This makes high pressure. Winds blow from high pressure to low pressure.

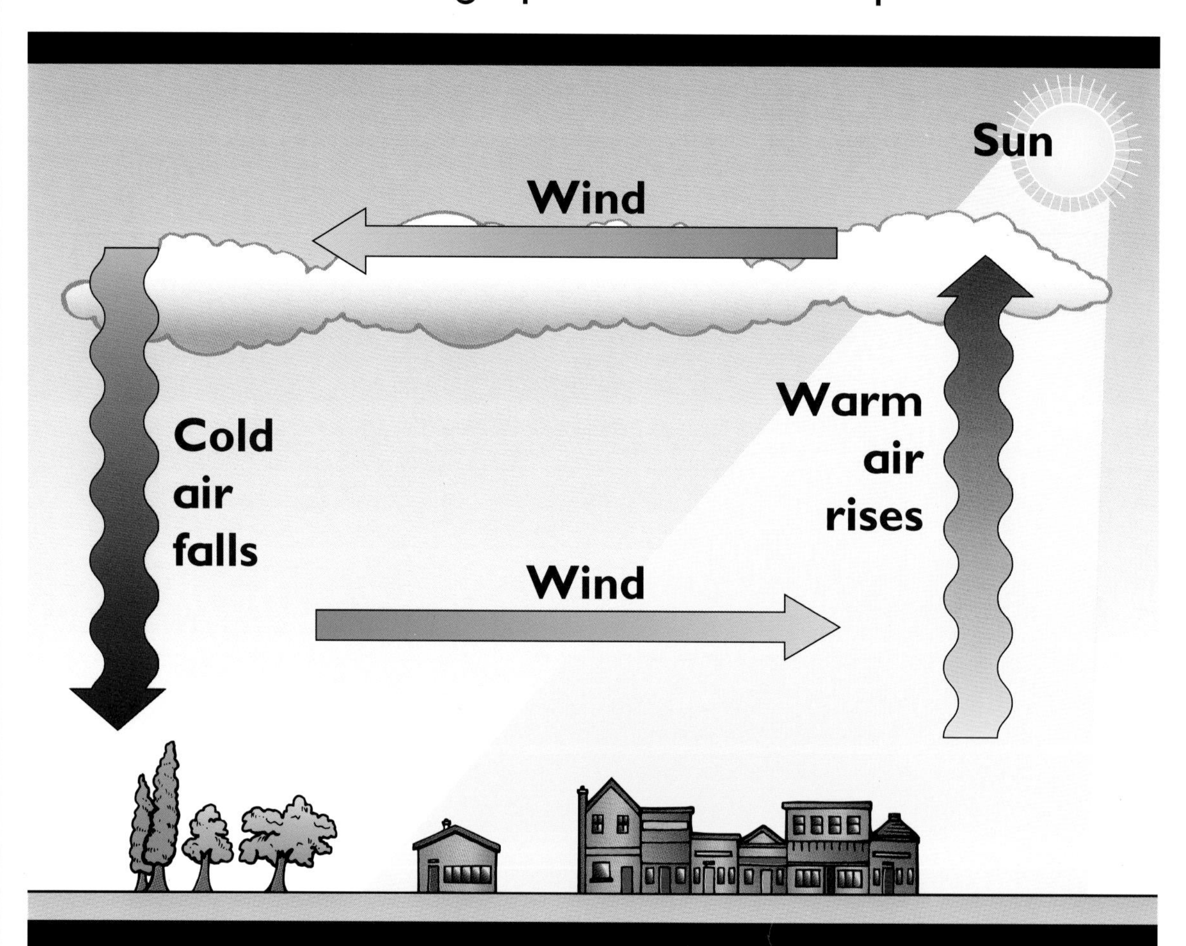

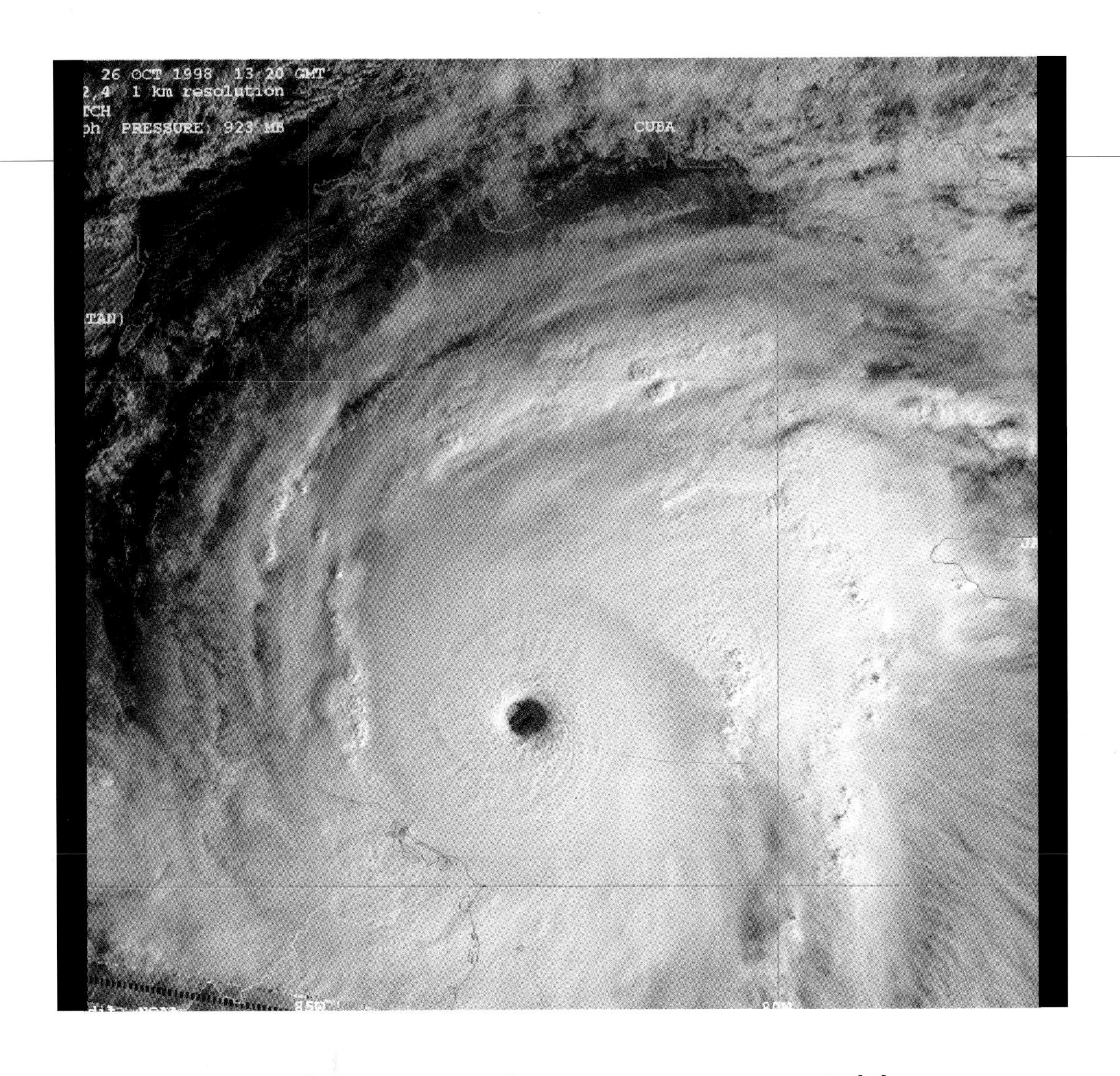

Hurricanes happen when air rises quickly over warm **oceans** in the **Tropics**. This makes a low pressure area. Strong winds rush in from high pressure areas. They form a **spiral** of wind and clouds around an area of still air.

What do hurricanes do?

Hurricanes form over the **ocean**. Hurricane winds are very powerful. The winds blow the sea into huge waves. Heavy rain falls from huge, dark clouds.

Hurricanes blow on to tropical **coasts**. As a hurricane blows over the land it gets less powerful. The hurricane can still do a lot of damage.

What are hurricanes like?

Hurricane winds blow through towns and cities. They uproot trees and blow the roofs off buildings. It is too windy for people to go outside.

Hurricanes bring heavy rain. They drop the water they have picked up from the **ocean** on to the land. This can often cause **floods**.

Harmful hurricanes

A hurricane **storm surge** has hit this **shore**. The surge is a huge wave pushed by the strong wind. The wave **floods** the shore. These boats have been carried on to the land.

Hurricanes damage roads, railways, bridges and airports. This makes it difficult for rescuers to reach people after a hurricane.

Hurricane in Jamaica

This is the island of Jamaica. It is a country that lies in the warm Caribbean Sea. People go there on holiday. Jamaica gets a lot of hurricanes.

This airport was damaged by Hurricane Gilbert. When airports are damaged, people cannot visit on holiday. This harms Jamaica's **tourist industry**.

Preparing for a hurricane

Weather forecasters can see a hurricane forming and moving in photos taken by **satellites** in space. The hurricane's cloud moves in a spiral over the Earth.

Television, radio and the Internet can warn people before the hurricane strikes. On the **coast**, officers from the **emergency services** use special flags to warn people that a hurricane is on the way.

Coping with hurricanes

People fix strong wooden boards to doors and windows before the hurricane strikes. This stops things blown by the hurricane from smashing the windows.

Many people go to community halls or special **shelters** before the hurricane hits. They are given food and a place to sleep. They stay until the hurricane is over and it is safe to go home.

Hurricane Andrew

Hurricane Andrew hit the United States in 1992. The hurricane battered Florida in the south of the country. People tried to prepare for it.

Hurricane Andrew damaged tall office blocks and an Air-Force base. The hurricane caused more damage than any other in the United States.

Nature and hurricanes

Hurricane winds and rain can destroy **crops**. In poor countries like Bangladesh these crops are very important. If they are destroyed, people will not have enough to eat.

Palm trees grow in the **Tropics**. These trees can bend and sway. This stops them from breaking in a hurricane.

To the rescue!

Some people cannot hide from hurricanes. Rescuers find people buried under fallen houses and trailers. Trailers get damaged easily by the strong winds.

Hurricane rains **flood** drains and water supplies. This makes drinking water dirty. Dirty drinking water can cause **disease**. So **aid workers** bring in fresh water as well as food.

Adapting to hurricanes

Belize is a country in Central America. A hurricane destroyed its **capital city** which lay on the **coast**. This picture shows the new city. It was built inland, well away from hurricanes.

Parts of Australia lie in a **hurricane zone**. People now have to build stronger buildings to protect them from hurricanes.

Fact file

◆ The worst known hurricane disaster happened in the country of Bangladesh in 1970. About 500,000 people died.

◆ There is a small area of still air in the centre of a hurricane. This place is called the 'eye' of the storm.

◆ Hurricanes are given names such as Shane or Tamsin. Their names go in alphabetical order. They go in order of boy and girl, too.

Glossary

aid workers workers who help people in a disaster
capital city most important city of any country
coasts where the land meets the sea
crops plants grown for food
disease illness
emergency services people who help us when there is a disaster. For example the police, ambulance and fire services.
flood overflowing water
hurricane zone area where hurricanes often happen
mass huge area or amount of something
oceans vast areas of sea
pressure pushing force
satellite spacecraft that moves around the Earth
shelters safe places
shore where the sea meets the land
spiral winding round and round
storm surge huge sea wave pushed to the shore by hurricane winds
tourist industry when people pay to visit places and attractions on holiday
Tropics very warm areas of the world on either side of the Equator. The Equator is an imaginary line around the fattest part of the Earth.
weather forecasters scientists who work out what weather we will get

Index